Messerschmitt Bf 109 G-6

Bf 109 G-6

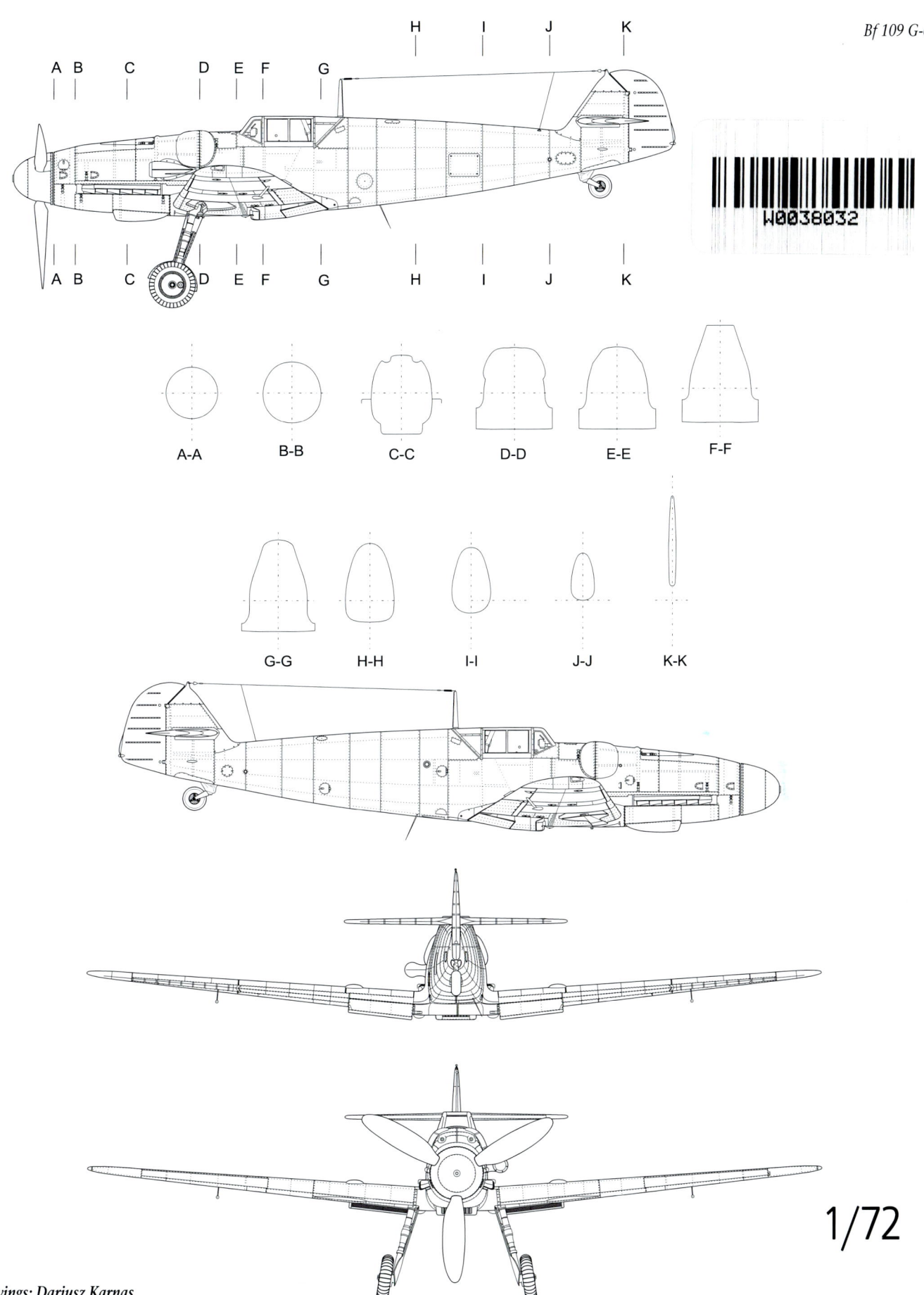

A B C D E F G H I J K

A-A B-B C-C D-D E-E F-F

G-G H-H I-I J-J K-K

1/72

Drawings: Dariusz Karnas

W0038032

1

Bf 109 G-6

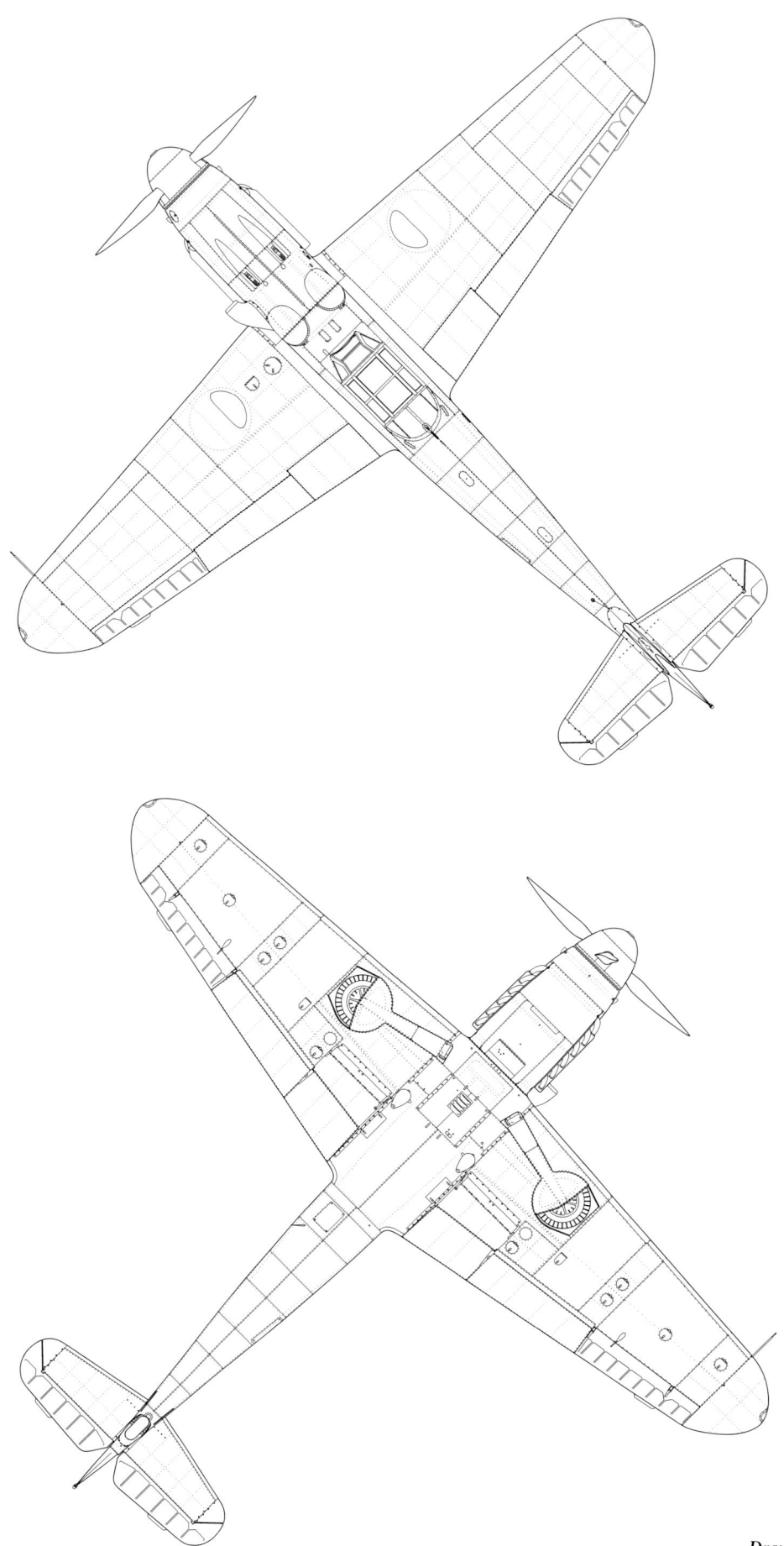

1/72

Drawings: Dariusz Karnas

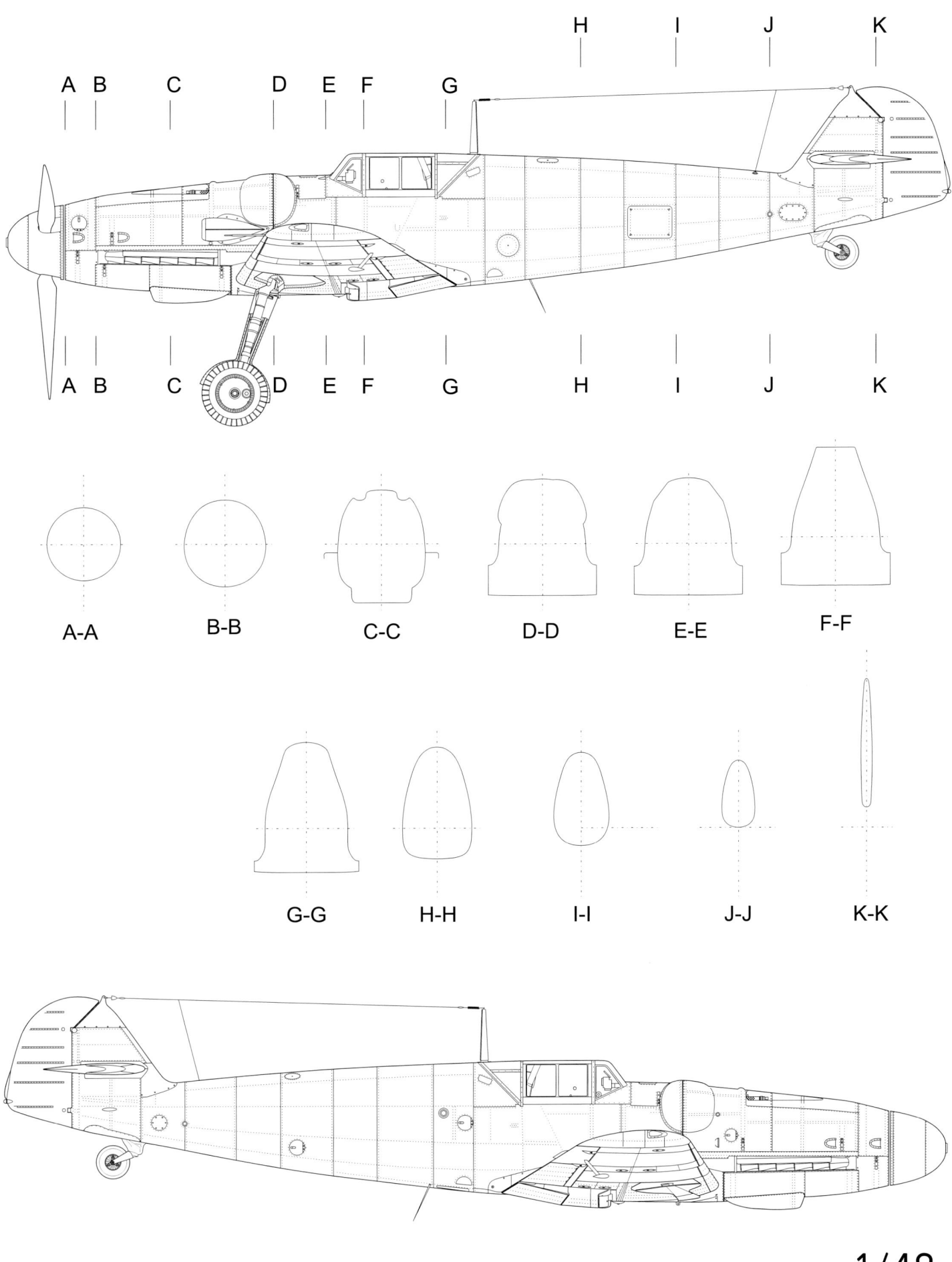

A-A B-B C-C D-D E-E F-F

G-G H-H I-I J-J K-K

1/48

Bf 109 G-6

Drawings: Dariusz Karnas

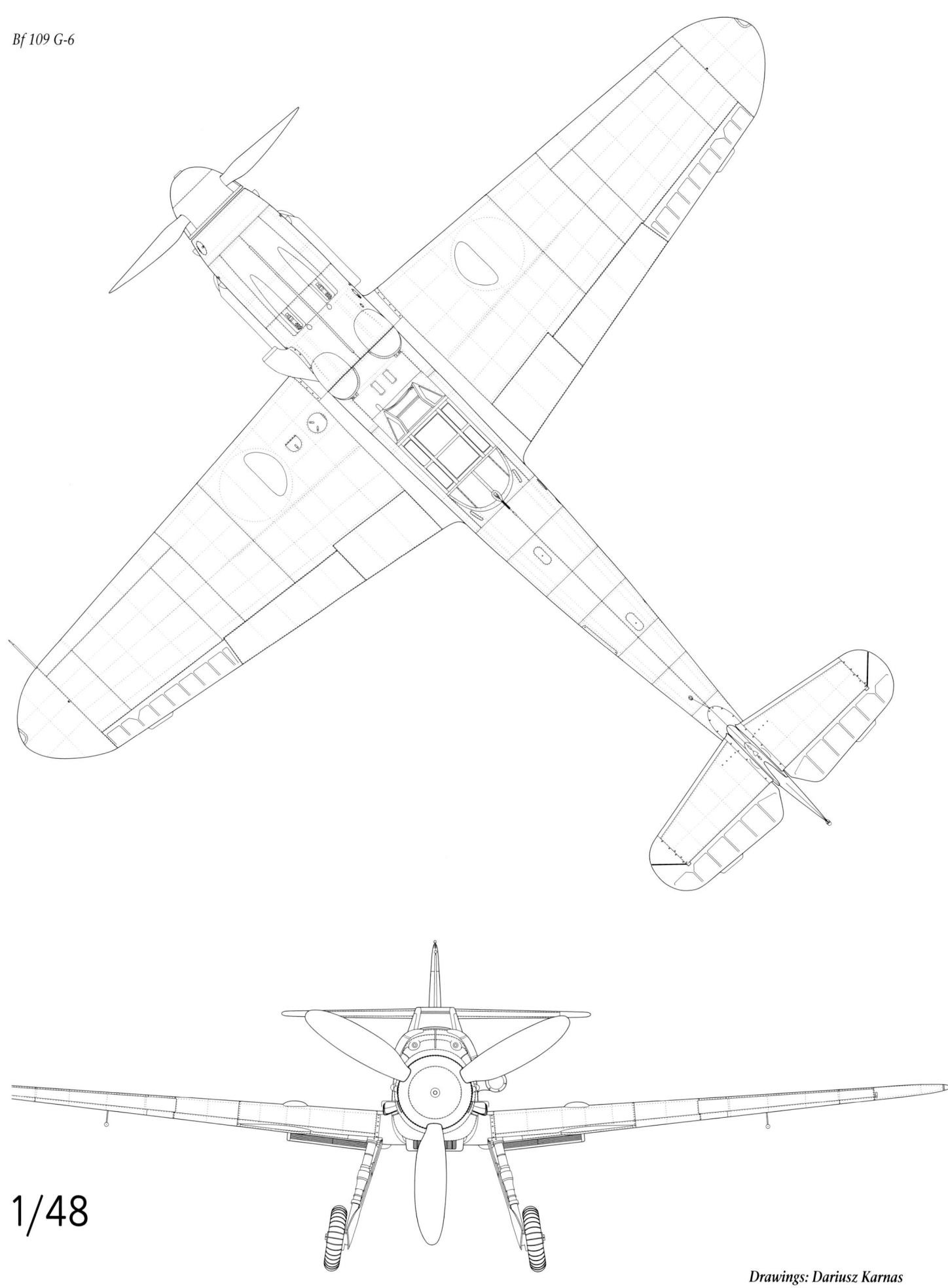

1/48

Drawings: Dariusz Karnas

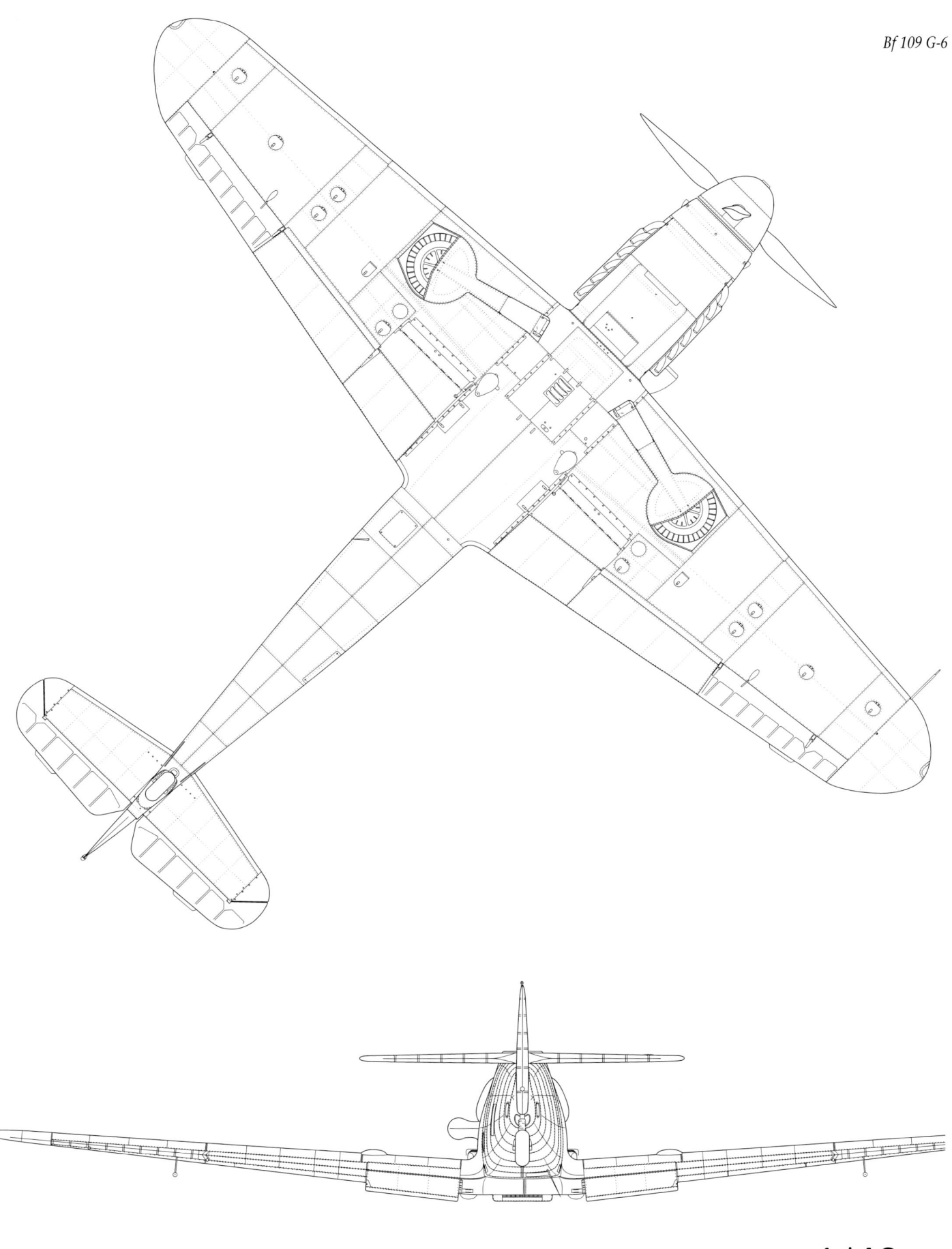

1/48

Drawings: Dariusz Karnas

Bf 109 G-6, White 13 of Grupul 9 Vânâtoare, Tecuci, Romania. Personal aircraft of Lt. Av. Tudor Greceanu, summer 1944. Royal Romanian Air Force. (Dénes Bernád coll.)

Bf 109 G-6/R6, (Rüstsatz VI), W.Nr. 15270, former "Gelbe 14" of 6./JG 53, captured at Comiso, Italy, summer 1943. Aircraft was sent to RAF Collyweston, 4 February 1944 and flown as VX 101. (US National Archives)

Cpt. Av. Şerbănescu Alexandru congratulated by his ground crewmen for a new victory, a Mustang shot down on 4 August 1944. (Dan Antoniu coll.)

9th FG pilots standing near Bf 109G-6 "Yellow 1", the mount of Grupul 9 Vânătoare commander Cpt. Av. Alexandru Şerbănescu. (Dan Antoniu coll.)

Bf 109 G-6, (Rüstsatz VI), White 7 of 4./JG 53. Flown by Lt. Günter Kremer. Wien-Seyring, early 1944. (Stratus coll.)

Bf 109 G-6, probably marked 'Red 26', standing at the edge of Kbely aerodrome. Aircraft probably of I./KG (J) 6. This aircraft had the VDM 9-12159A broad blade propeller. (via Bohumir Kudlička)

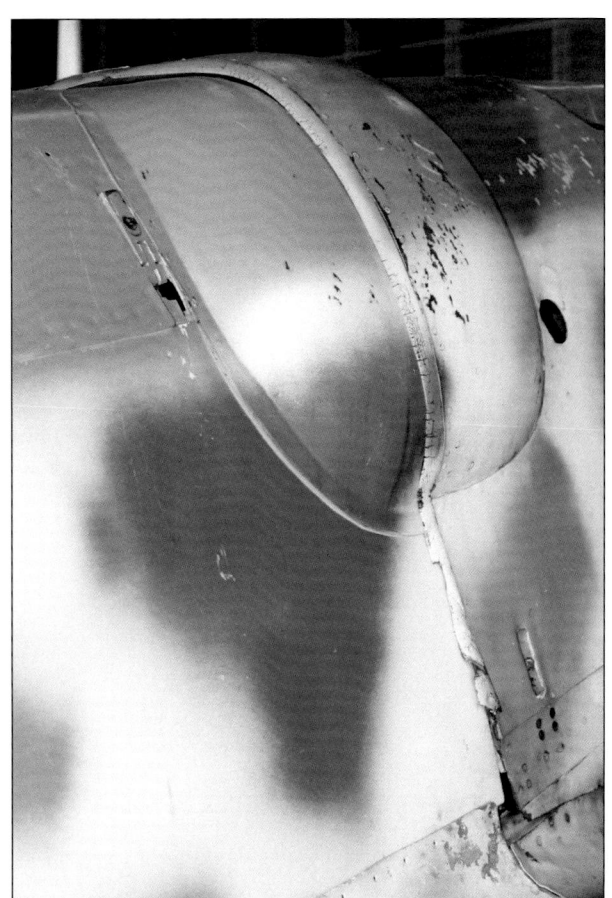

Above: Front fuselage inlet and front lever lock.

Above right and below: Two photos of the fuselage bulge typical for G-6 and G-14 versions. (All photos R. Panek)

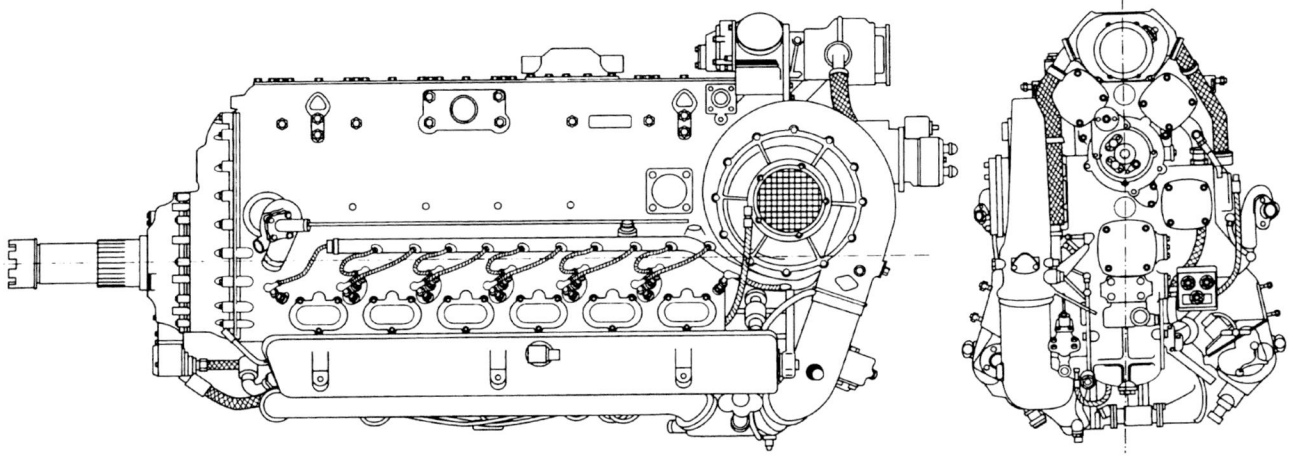

Top: DB 605 A engine mounted in Bf 109 G-6.
(M. Tyburski)

Above: Port and rear views of the DB 605 A engine. Drawings not to scale.

Left: DB 605 A engine preserved at National Air and Space Museum, Washington, D.C. USA.
(R. Pęczkowski)

DB 605 A engine overhaul in the field. Note cowling removed. (Dénes Bernád coll.)

Bf 109 G-6 under restoration. DB 605 A engine have just been mounted still without its wiring and other equipment. (D. Karnas)

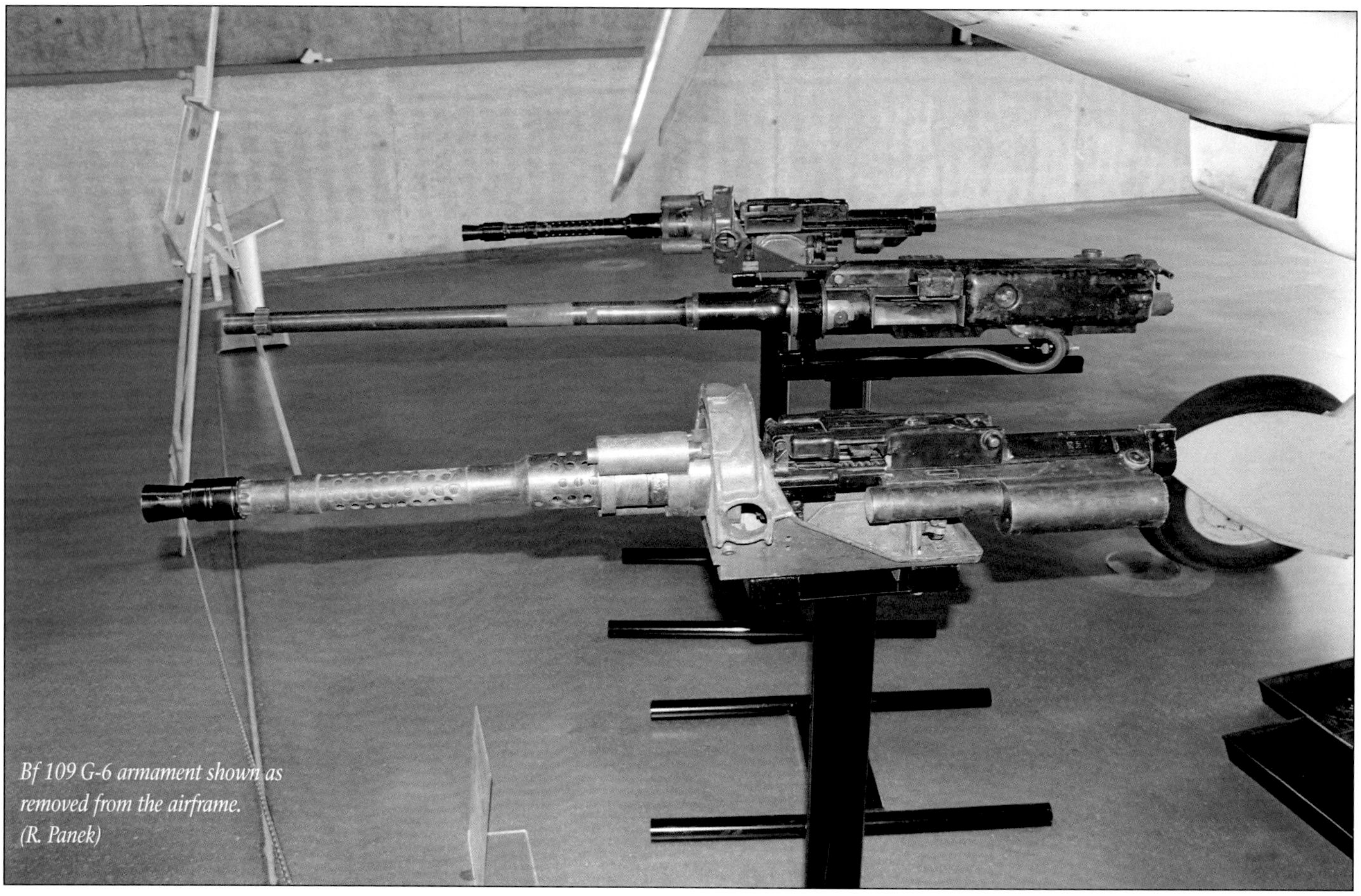

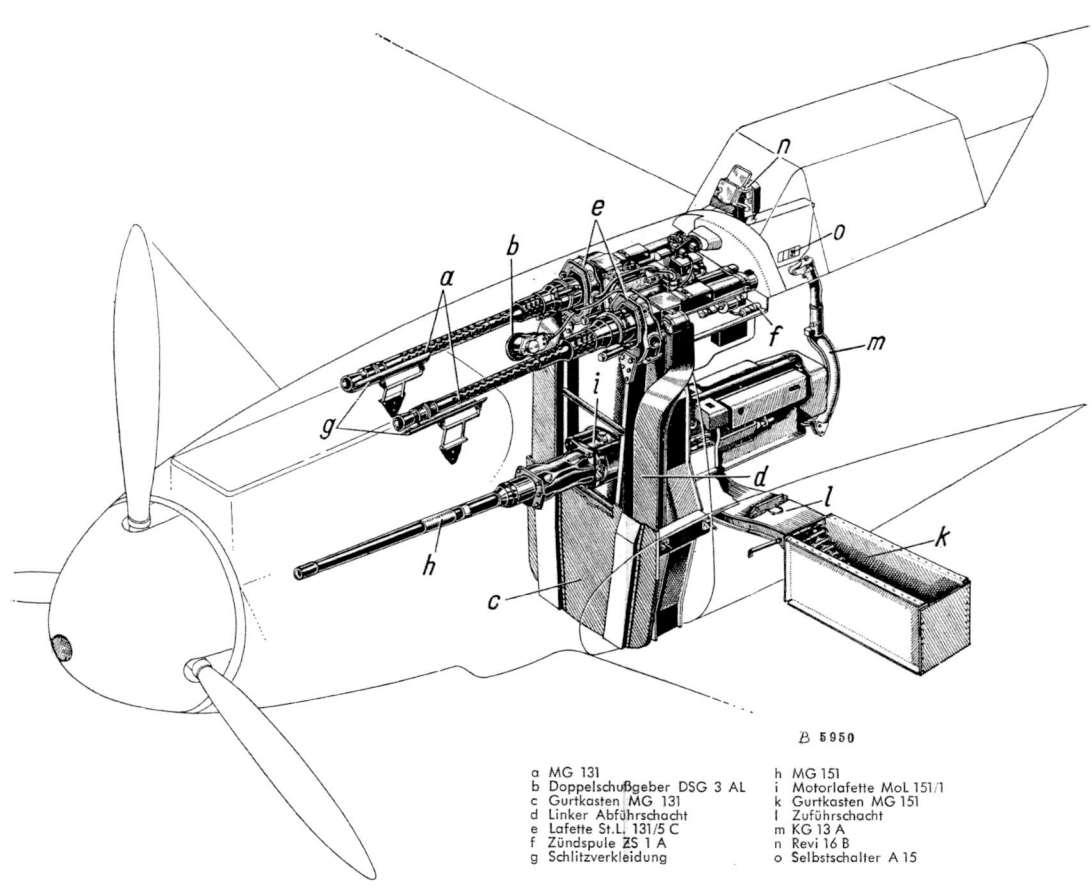

Bf 109 G-6 armament shown as removed from the airframe. (R. Panek)

Below: *Drawing from the Handbuch, showing the standard G-6 armament.*
Selected symbols: a. MG 131 guns; c. MG 131 ammunition container; d. MG 131 case chutes; k. MG 151/20 cannon ammunition; container; m. KG 13A control stick; n. Revi 16B gun sight. (Technical Manual)

a
b
c
d
e
f
g
h
i
k
l
m
n
o
B 5950
a MG 131
b Doppelschußgeber DSG 3 AL
c Gurtkasten MG 131
d Linker Abführschacht
e Lafette St.L. 131/5 C
f Zündspule ZS 1 A
g Schlitzverkleidung
h MG 151
i Motorlafette MoL 151/1
k Gurtkasten MG 151
l Zuführschacht
m KG 13 A
n Revi 16 B
o Selbstschalter A 15

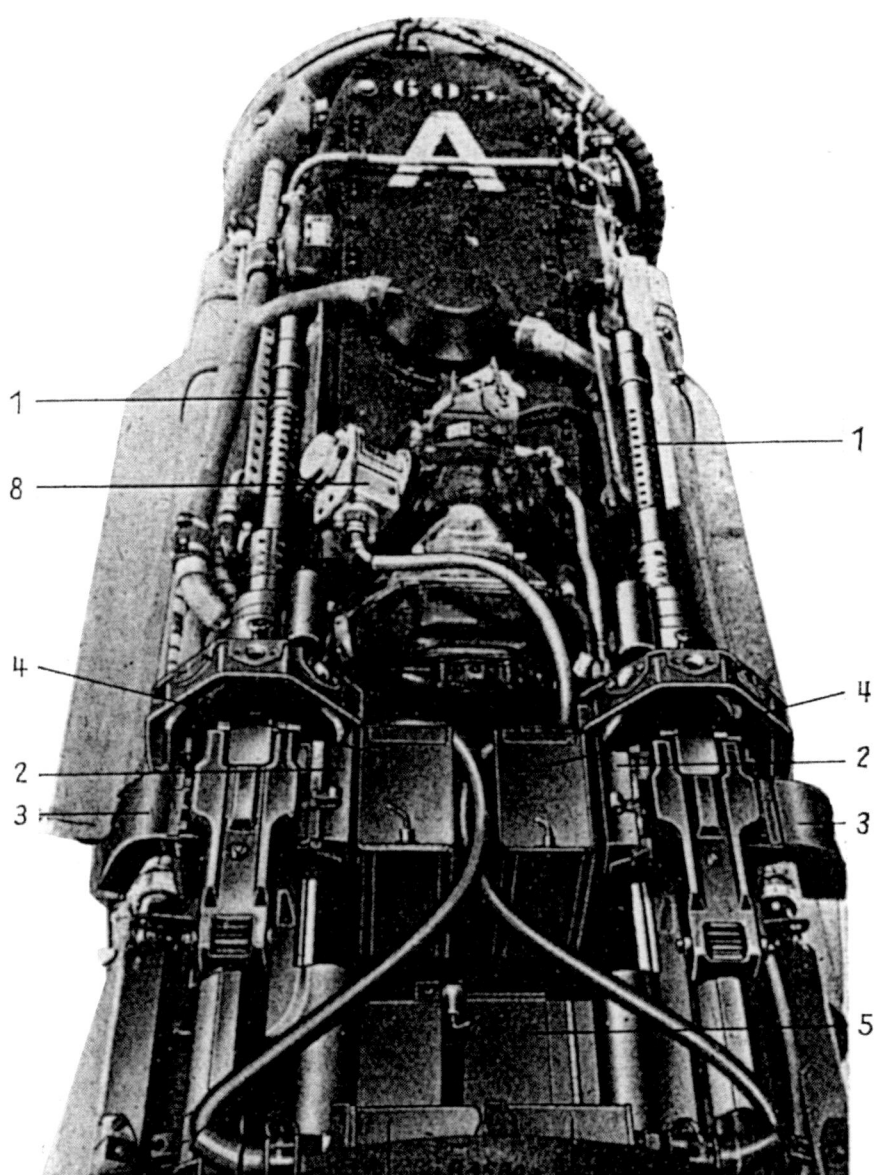

Above: MG 151 cannon fairing in the G-6 cockpit.

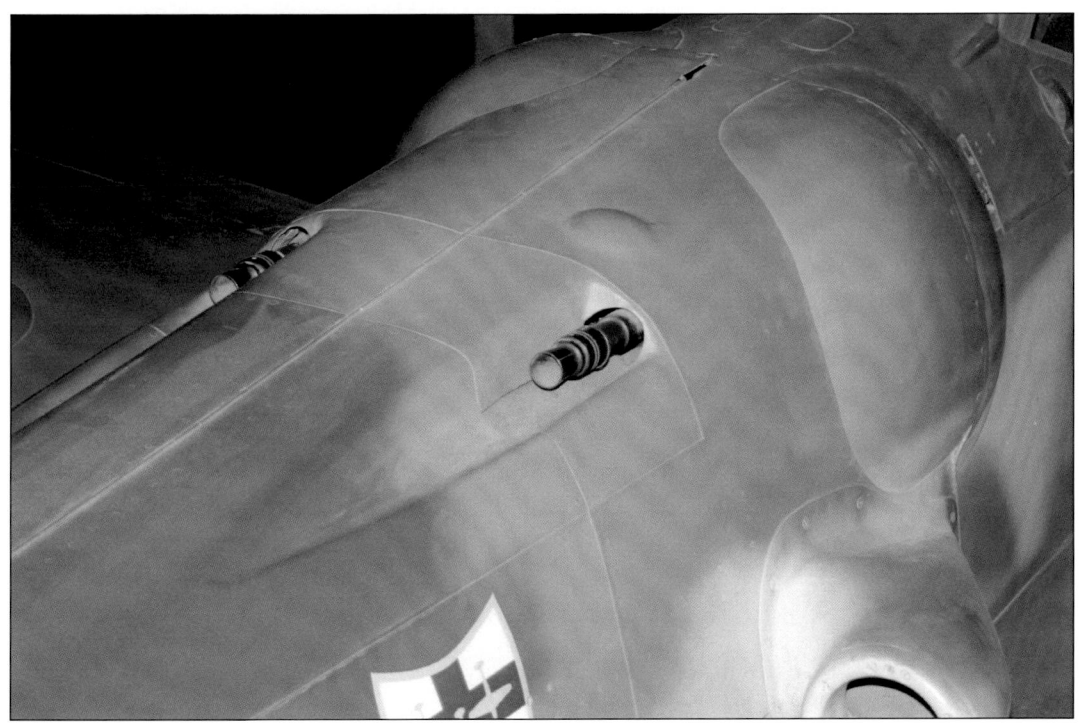

MG 131 machine guns in the Bf 109 G-6 fuselage.
(R. Pęczkowski)

Starboard, main undercarriage leg of the G-6 version. Note later style wheels. (R. Panek)

Below, right: *Starboard undercarriage well. (R. Panek)*

Main wheel details. Later style as was used from G-4 version. Sometimes seen also on the late G-2. (R. Panek)

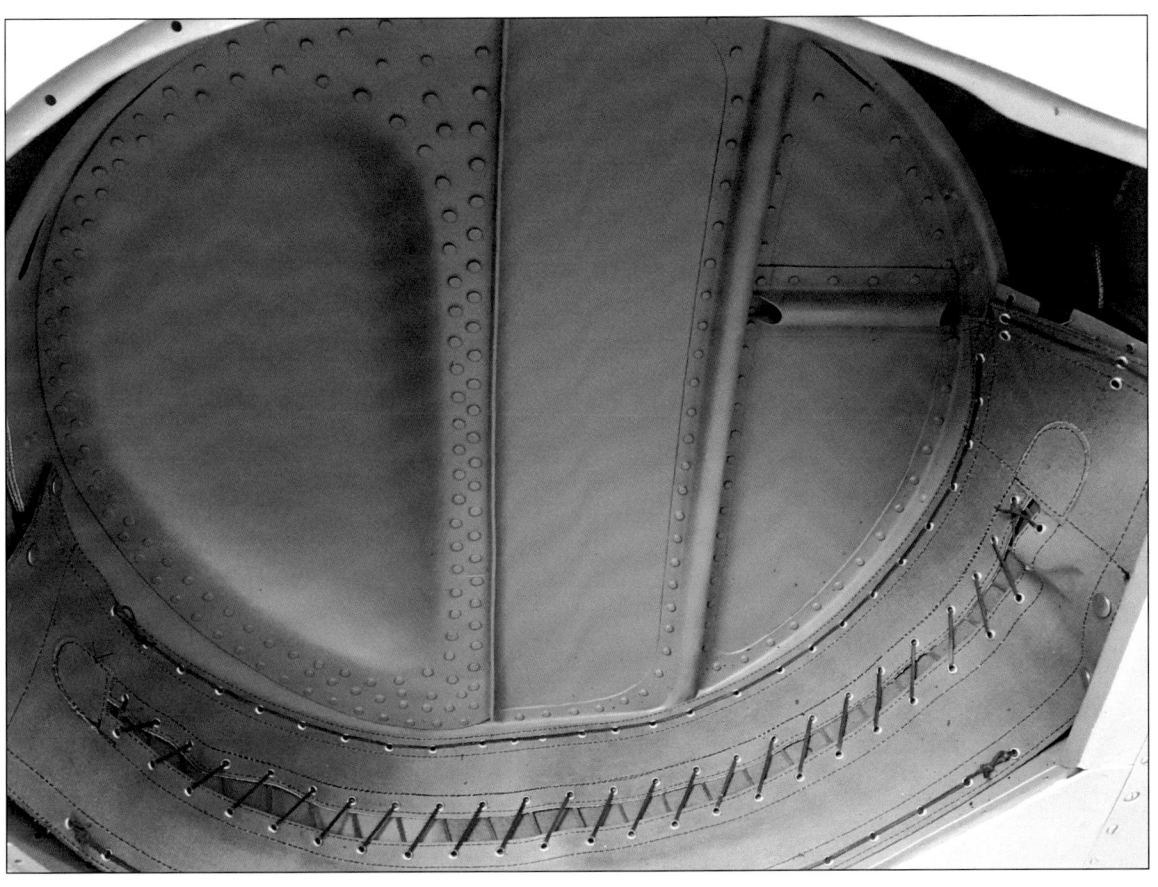

Main wheel well with the bulge. Bf 109 G. (S. Myers)

Left: *Bf 109 G-5 sporting its starboard main undercarriage leg and wheel. Note additional fuel tank. (E. Všetečka)*

Below: *Similar photo, but of Bf 109 G-6. (R. Panek)*

Bf 109 G tail wheel.
(D. Lukić)

Below: *Tail wheel construction shown in Spare Parts Catalogue.*

Bottom, left: *Port side of the tail wheel. (S. Myers)*

Bottom, right: *Tail wheel shown from the rear.*
(R. Panek)

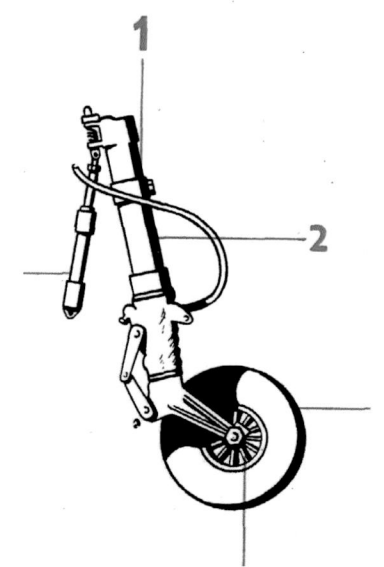

Upper view of the G-6 canopy. Note rear armoured glass. (A. Juszczak)

Rear armoured glass seen from the right. (Stratus coll.)

Bf 109 G-6 canopy, port side. Note the side window in fully open position. (M. Tyburski)

Bf 109 G-6 windscreen details. Revi gun sight is also visible. (Stratus coll.)

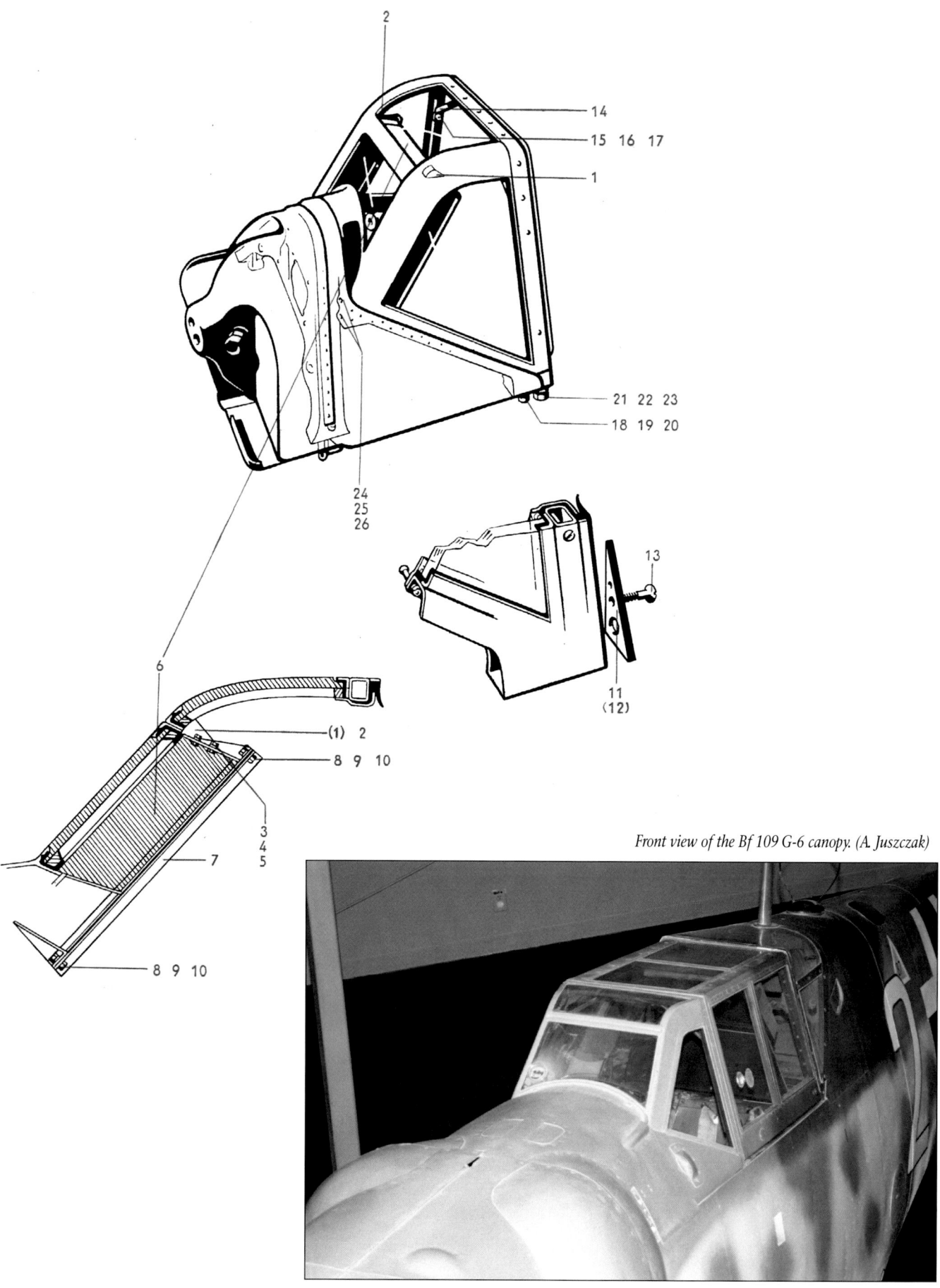

Drawings of the Bf 109 G windscreen. (Spare Parts Manual)

Front view of the Bf 109 G-6 canopy. (A. Juszczak)

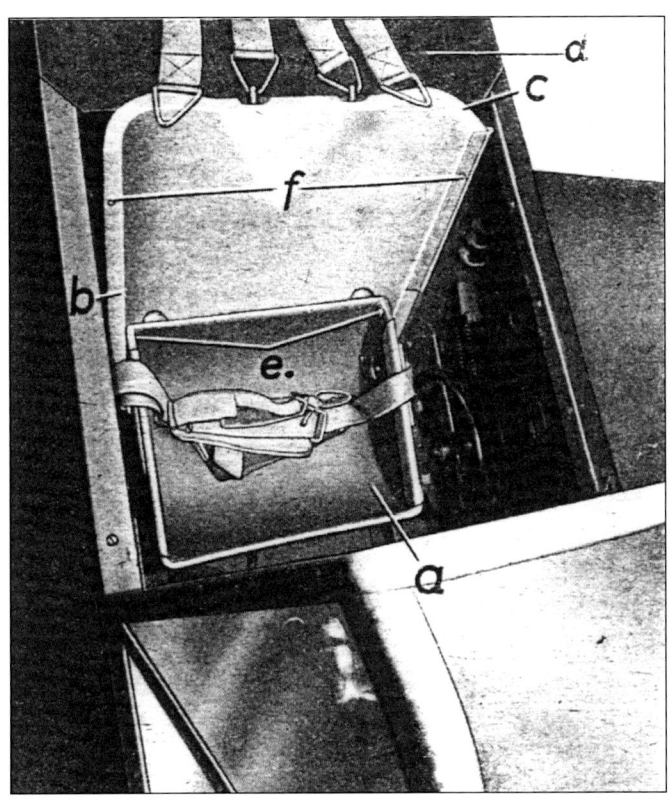

Pilot's seat in the G-2. (Technical Manual)

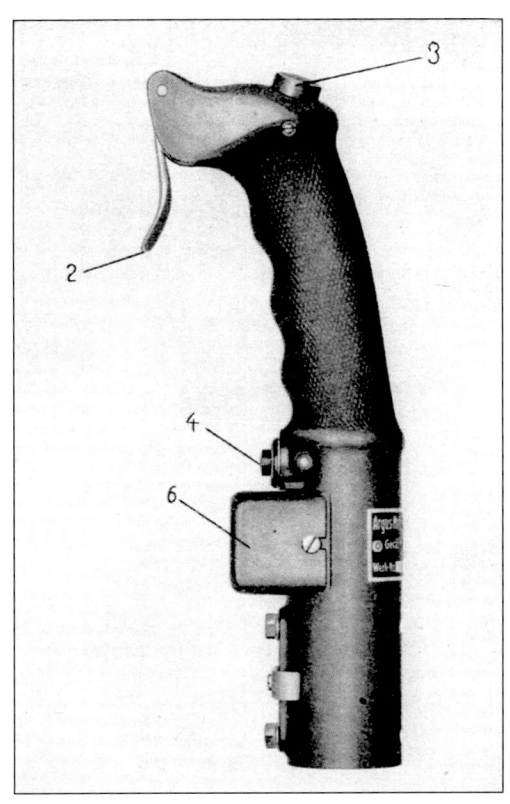

Early Gustav pilot's control column – KG 12 type. (Technical Manual)

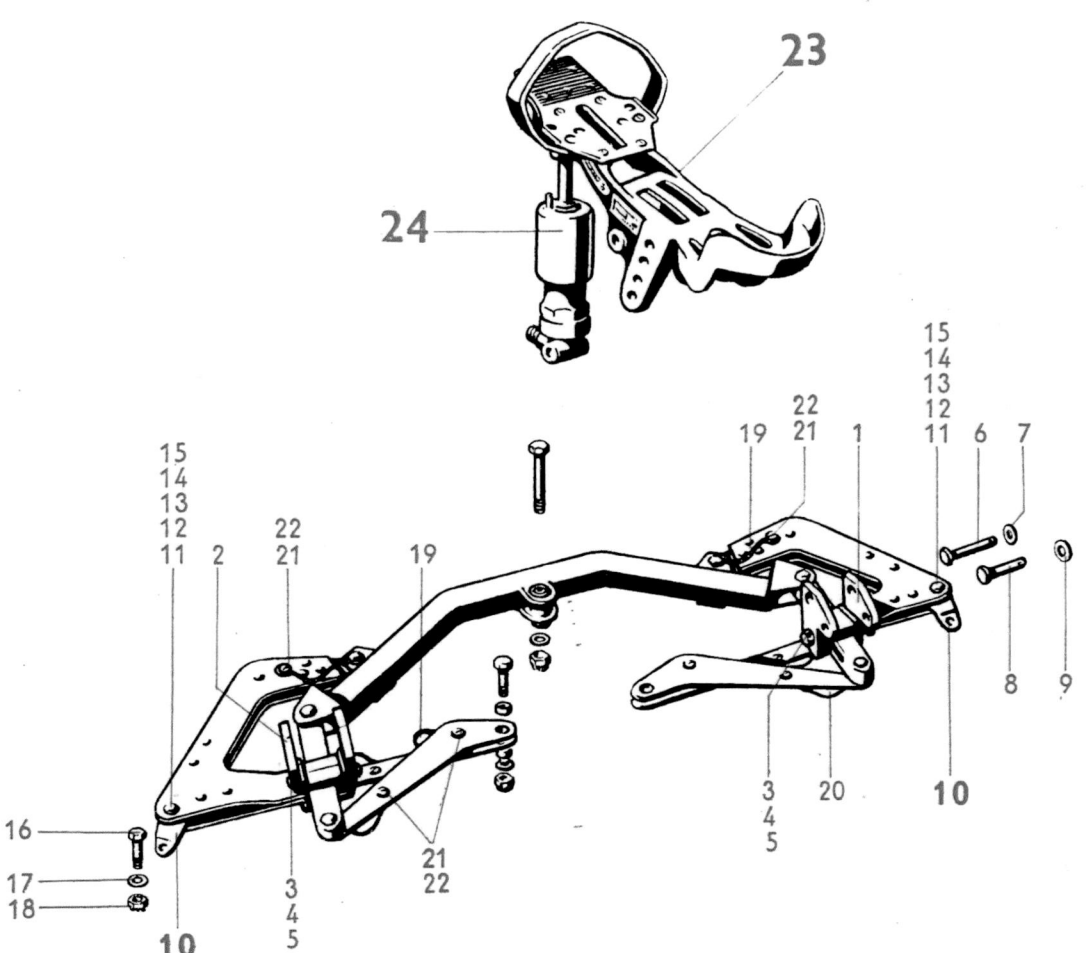

Details of the rudder pedal. (Spare Parts Catalogue)

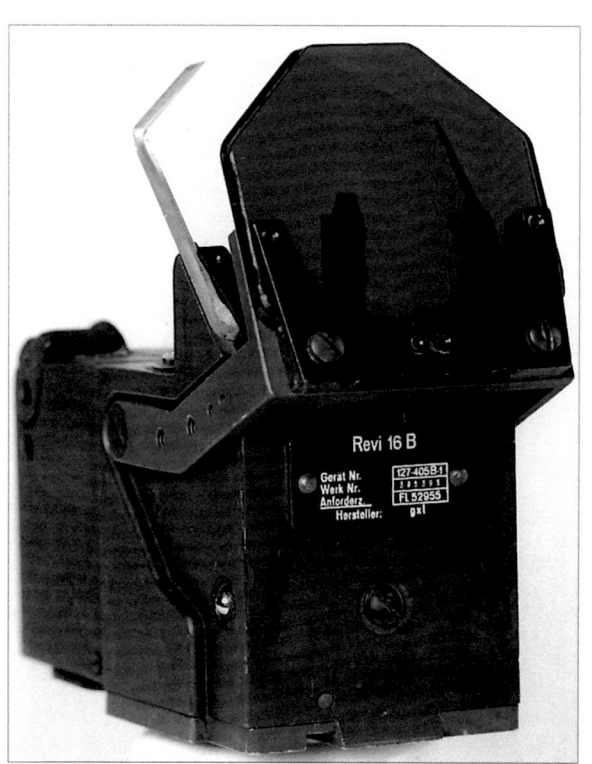

Flaps and fin adjustment wheels mounted on the port cockpit wall.
a. wheels;
b. fuselage stringer;
c. mount.
(Technical Manual)

Revi 16B gun sight. (Technical Manual)

Revi 16B 3D drawing. (D. Grzywacz)

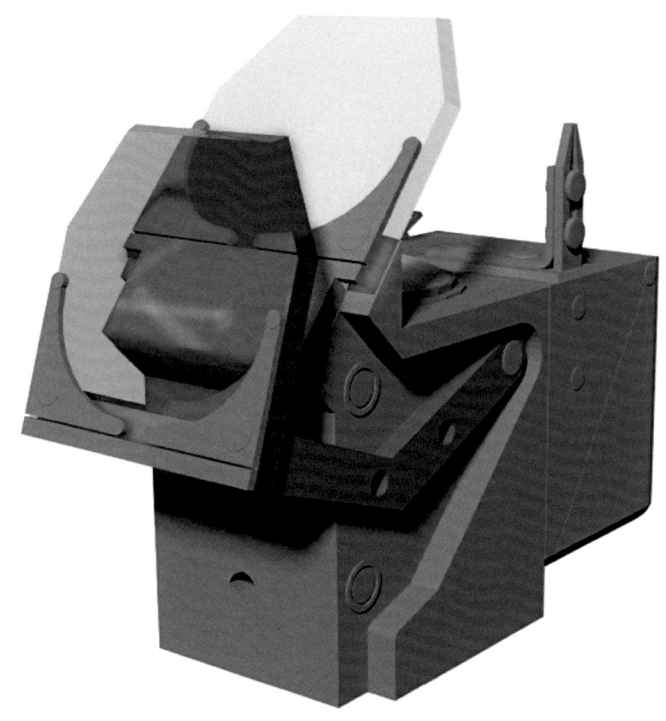

Bf 109 G-6 instrument panel.
(Drawing by Dariusz Karnas)

1. Port under-wing cannon control lamp;
2. Fuselage gun round counters;
3. Board clock;
4. Starboard under-wing cannon control lamp;
5. Ignition switch for magnetos panel (Zündschalter, Fl 21121-2);
6. Engine starter switch (Netzausschalter, Fl 32315-2);
7. Repeater compass (Führertochterkompass, Fl. 23334);
8. Artificial horizon combined with turn and bank indicator (Wendehorizont, Fl.22410);
9. Manifold boost pressure gauge (Ladedruckmesser, Fl.20555)
10. AFN indicator (Anzeigegerät für Funknavigation AFN 2, Fl 27002);
11, 12. Electrical switches

13. Altimeter (Fein-Grobhöhenmesser, Fl. 22322);
14. Airspeed indicator (Fahrtmesser, Fl. 22234);
15. Tachometer (Drehzahlmesser, Fl.20222-3);
16. Propeller pitch indicator (Mechanischer Luftschrauben – Stellungsanzeiger mit Zentral – Anschluss, Fl.18503-2);
17. Oil temperature gauge (Schmierstoff Temperaturanzeiger, Fl.20342-2);
18. Undercarriage position indicator (Fahrwerksanzeige – Vierlampengerä, Fl 32526);
19. Fuel level warning lamp (Reststandswarnlampe Fl 32262-1);
20. Undercarriage emergency control lever;
21. Fuel indicator (Kraftstoffvorratsanzeiger, Fl.20723)
22. Fuel and oil pressure gauge (Doppeldruckmesser, Fl.20512-1)

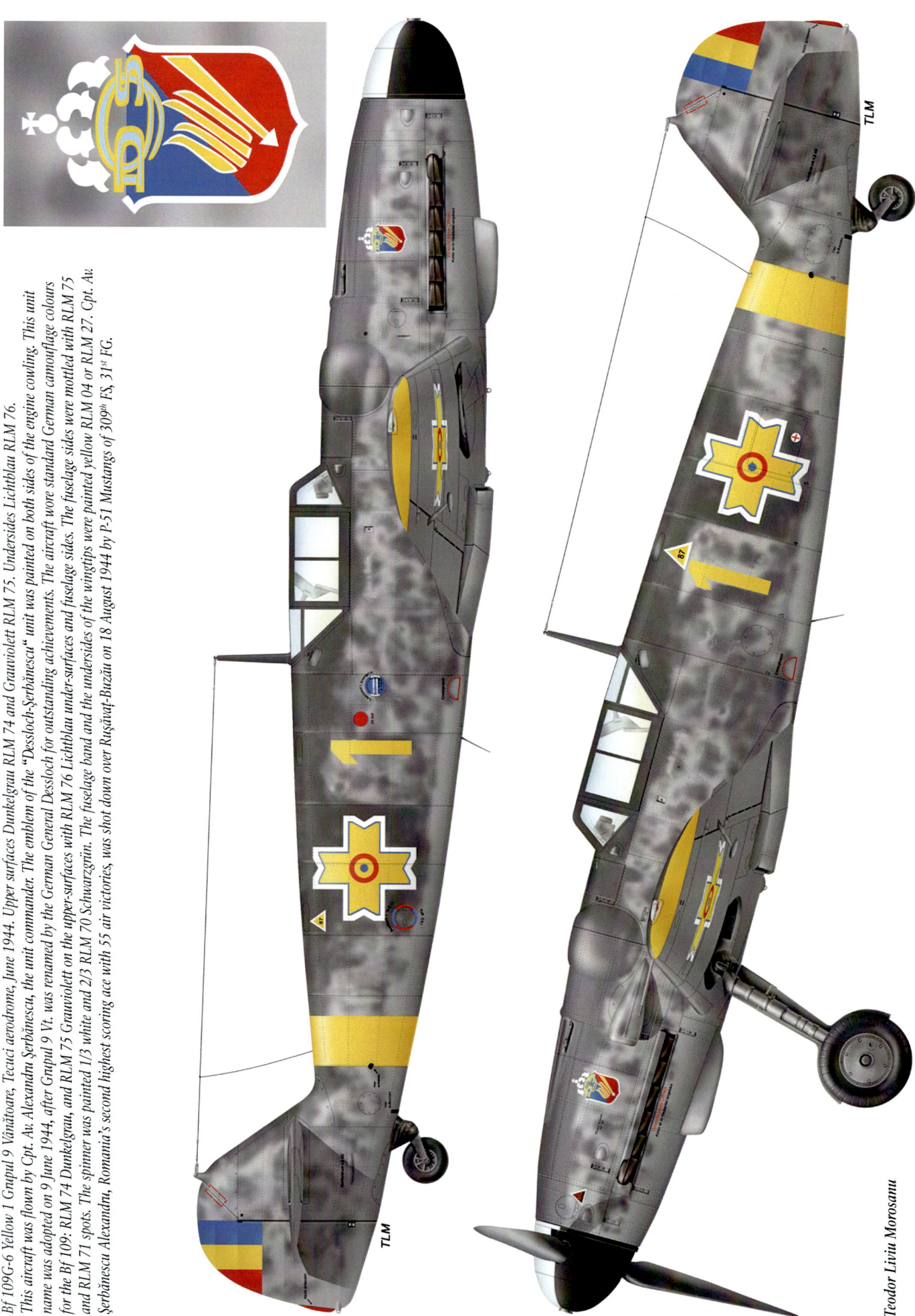

Bf 109G-6 Yellow 1 Grupul 9 Vânătoare, Tecuci aerodrome, June 1944. Upper surfaces Dunkelgrau RLM 74 and Grauviolett RLM 75. Undersides Lichtblau RLM 76.

This aircraft was flown by Cpt. Av. Alexandru Şerbănescu, the unit commander. The emblem of the "Dessloch-Şerbănescu" unit was painted on both sides of the engine cowling. This unit name was adopted on 9 June 1944, after Grupul 9 Vt. was renamed by the German General Dessloch for outstanding achievements. The aircraft wore standard German camouflage colours for the Bf 109: RLM 74 Dunkelgrau, and RLM 75 Grauviolett on the upper-surfaces with RLM 76 Lichtblau under-surfaces and fuselage sides. The fuselage sides were mottled with RLM 75 and RLM 71 spots. The spinner was painted 1/3 white and 2/3 RLM 70 Schwarzgrün. The fuselage band and the undersides of the wingtips were painted yellow RLM 04 or RLM 27. Cpt. Av. Şerbănescu Alexandru, Romania's second highest scoring ace with 55 air victories, was shot down over Rușavaț-Buzău on 18 August 1944 by P-51 Mustangs of 309th FS, 31st FG.

Teodor Liviu Morosanu

22

Bf 109G-6 Yellow 1 Grupul 9 Vânătoare, Tecuci aerodrome, June 1944.

Bf 109G-6 Yellow 1 Grupul 9 Vânătoare, Tecuci aerodrome, June 1944.

Teodor Liviu Morosanu

24